An HOLISTIC Way In Saving The "Honeybee"

by
John Harding

Published in the United Kingdom by
Northern Bee Books, Scout Bottom Farm,
Mytholmroyd, West Yorkshire HX7 5JS
Tel: 01422 882751   Fax: 01422 886157
www.GroovyCart.co.uk/beebooks

ISBN 978-1-904846-70-3

© An HOLISTIC Way In Saving The "Honeybee"
by John Harding

First Published 2010

All photographs taken are by the author

Designed by D&P Design and Print
Printed by Lightning Source UK

An HOLISTIC Way In Saving The "Honeybee"

by
John Harding

An HOLISTIC Way In Saving The "Honeybee"

With

A Natural Phenomenon!

"Electromagnetic Geopathic Stress Lines"

Incorporating

"The Harding Queen Rearing System Using Two Queens"

And More

"The Holy Grail of Beekeeping Has Been Found!"

By

John Harding

"Read What You See In Your Beehive"

"Don't See What You Have Read, In A Book!"

FORWARD

Dear Reader.

I am sure you are aware of the plight of the Honeybee worldwide. Beekeepers need an answer. Initially Apiarist worldwide was putting the blame for Honeybee demise on the doorstep of the Chemical and Mobile Phone Industries.

Honeybees are dying out at an alarming rate with no one knowing why. Pesticides, CCD, GM crops, Climate change, Mobiles, Global warming or perhaps someone or something to blame would be acceptable to everyone. There are many possibilities being put forward but as yet, no answers. The Varroa mite is not helping matters with its contribution.

However there are two common denominators why Honeybees are dying worldwide. A short explanation first.

Chemical companies are investing millions worldwide in Universities, Scientists, Professors, Doctors, Institutes, Beekeeping Organisations and whoever, so they just might find a *chemical or bacterial* answer for the parasitic mite called Varroa that is sweeping the continents devastating Honeybees. Mobile Phone Companies are in denial not wanting the blame.

Chemical companies need an answer whether it is one or the other so they may recoup their investment and profit from beekeepers worldwide in selling their product.

Was Einstein right in his alleged statement? *"If Honeybees die out then mankind will follow 4 years later"* the chances are that it won't be 4 years due to other foods such as rice being available but it will happen eventually as honeybees do pollinate 35% of what we eat.

Once Honeybees are gone, Honeybees are gone for good!

I am a beekeeper of 30 years` experience, keeping up to 300 beehives, until 6 years ago. I have invented beekeeping equipment in that time that I am proud to say, does bear my name, *"The Harding Queen Rearing System using Two Queens"* and *"The Harding Mini Nucleus Complete System"* (as seen on the internet website for BIBBA). These are an inclusion of this book, Chapter Three & Five. During my life's work things happen and you wonder at nature, how perfect is the Honeybee micro-world, why would you want to change it and yet mankind unknowingly has changed the Honeybees perfect 200 million year existence to what mankind wants.

My beekeeping puzzle is based on observation and logic over the past 30 years with each piece complimenting the next, eventually creating a picture and discovering;

"The answer and solution to the Holy Grail of beekeeping".

I have always thought there was a natural way to treat the parasitic mite Varroa. After 18 years without treatment of any chemicals or sugar in my hives I have found the answer and it is a *"World Exclusive!"* It didn't start with the Varroa mite 20 years ago, what the Varroa mite did was escalate the problem to what beekeepers had done worldwide, but it did bring it to the attention of the media and mainstream public in the last few years causing an over re-action due to Einstein's alleged quote.

Honeybees started dying out when man found honey, thousands of years ago when man wanted to domesticate Honeybees to harvest honey, putting them into logs, boxes, skeps eventually beehives but taking them away from their natural source of survival and requirements, which keeps their delicate micro-environment alive.

The first common denominator for the demise of Honeybees is…
Mankind! Well, Beekeepers now and in the past!

So what is the second common denominator?

"I have found a natural phenomenon, the bees need it to survive to complete their micro-existent world, and is free. I am the first person in the world to combine Honeybees with this phenomenon, so you can imagine how the chemical companies are going to react after spending millions around the globe. I have approached Universities and Beekeeping Organisations here, in the UK, and abroad with my hypothesis but due to the infiltration of funding from chemical companies or others, University Scientist, Professors or Scholars are unable to take my hypothesis due to inevitably losing their precious funding then being biased to a chemical or bacterial answer".

Yes! It is topical, political and controversial! One single person taking on the might of a billion-dollar pharmaceutical industry and the Hierarchy of the Beekeeping World with every beekeeper past and present being the reason for their demise and the answer being a natural phenomenon which is free.

CCD (Colony Collapse Disorder) in the USA is also down to Mankind for the demise of their Honeybees having the same problems as us but with one extra reason that is *only* in the USA.

Whatever you think after you have read this, I will not be popular with any beekeeper, scientist, professor or anyone looking for a chemical or bacterial answer, but may, just may, stop Honeybees dying out worldwide. That will be pleasing in itself. I am just a passionate beekeeper that has found an answer and solution. This book is a small part of the invisible world of the mysterious Honeybee that is disappearing too quickly.

Please enjoy.

CONTENTS

Acknowledgments

Doctor Brian Jackson

He brought his tortoise to me for an MOT as I run a rescue centre for tortoises and we became friends, so.

For his belief and inspiration to get me to type one word and the rest would follow, becoming this book, he was right (tortoise book next).

Stuart Bailey
Chairman of Rowse Honey

For listening, encouragement and belief, also allowing me to use his name as an introduction.

Geoff Tristram

For his candid truthful remarks and advice albeit blunt but funny.

Rolf Gordon

For his insight into Cancer Research.

Thanks must go to...

My ex wife Cynthia *(Who I will Love forever)*
If she had not have left, I would never have found the confirmation to the answer and solution

My Children
Christopher & Sally and their partners Rachael & Raphael

My Grandchildren
Tahlia, Oliver & Luca

Alan Bishop
My mentor when I started this mad journey, God rest his soul, thank you Alan you will always be remembered

Tom Bryson
For his help, guidance and belief at the beginning, being the first to read the manuscript when in its infancy

My drinking pals @ The Bathams, Kinver
For without their sarcasm this book would never have been done

Nicky
It was her cancer leaflet that I read, thank you Nic

PHOTOGRAPHS

INTRODUCTION

How did all this come about?

After I finished racing motorbikes for 12 years, racing at all major circuits in the UK, stamp collecting did not create the same adrenalin until a friend of mine invited me over to see his beehives and when he opened a full size colony, WOW!!! The adrenalin was oozing! This was for me so within 10 days a beehive was in my garden....The rest is history....

I feel that after reading every honeybee magazine that I receive, I should be a politician, scientist, doctor or pharmacist due to the complexity of how this craft is evolving. Yes I know it is a very diverse subject and due to politics changing farming by people in suit's that possibly has never set foot on a farm in this country has changed the countryside from what it was 50/60/70 years ago.

However, is it for the better?...

Is it just me that thinks that, as a full time beekeeper?

I love practical beekeeping. This involves helping the bees to overcome any drawbacks as this enables them to maximise on their health and increases the resulting honey crop. This brings with it, being a part time botanist, weather reporter and much more including meeting all classes and backgrounds of people which are very diverse from the very rich to the very poor and they allow you into their lives and homes, some Stately, due to you knowing about the mysterious Honeybee.

However, and it will never cease to amaze me as to their first question always being the same "Do you ever get stung?" I can never understand why they ask this as I thought it was pretty obvious. My reply is however somewhat left to their imagination as my description does include "just feeling a little prick" which normally brings a smile even to the most Victorian of person. It certainly does break the ice! You never know where beekeeping is going to take you as every day is so different from the last. It is definitely not a 9 to 5 job it is hard work but very satisfying and rewarding.

What could be more perfect than beekeeping?

I now feel the time is right to tell others of the system that includes my designs, research, findings and experiments. I have always been concerned, after telling many of my findings here and abroad, that someone else will capitalise on them before I publicise them myself. Some ideas are not new; it is just the combination of them that makes everything work.

Being a master of beekeeping does not necessarily mean being able to take as many exams as you can or how many books you can write or even as a Scientific Professor of Entomology that can write volumes of text. The honeybees do not care about how clever you are. It is how gentle you handle and manipulate your bees, your decisions of what you see and do, and how you prepare for the next inspection. All the bees want is quite simple; *"it is reading what they would require if they were in their wild environment"*, however the problem is you have to read that situation. It is not the bees fault if it goes wrong, it is yours for not reading what you have seen.

They did not ask to be put in a beehive, it is what mankind wanted, honeybees will never be domesticated and will always be a wild part of nature.

I am neither academic nor scientific but more practical in my approach to beekeeping. It is very much what the heading says;

"Read what you see, in your beehive, don't see what you have read about, that should be in the beehive!"

If you can do that then you are starting too learn the art of what the bees want, beginning your journey to become a Master Beekeeper in understanding the mysterious Honeybee.

I have been appalled at how often I have witnessed beekeepers, old, experienced, new and from the Ministry, who badly handle the colonies; some with very little consideration for the bees at all, crushing many bees while inspecting, just because the beekeeper has protection, then all respect for the bee disappears and they wonder why they get stung or the bees are aggressive.

Poor handling can then spiral into fear and bad management where colonies are not dealt with when they most need it, causing the colony too swarm, virgin queens failing to get mated or mated with the wrong strain, leaving a nasty temper hive that is a nuisance to everyone before it eventually dies out or is dealt with.

Whenever I am approached by someone that wants to start beekeeping I tend *NOT* to go down the normal route of most beekeepers do in saying;

"Come and get stung, let's see what your reaction is and if all is well start them off with a nucleus or a swarm so as to break them in gently while training".

A newcomer should know what they are letting themselves in for; so I get them stung as normal but then take them to a full size colony that has 60,000 plus bees in. It is now at this point that this person will change his/her mind very quickly and realise what beekeeping is all about in coming to terms with *"IT IS"* or *"IT IS NOT FOR THEM"* and asking themselves an awful lot of questions in a very short space of time, such as; "I was intending to put a hive in my postage size stamp garden or putting a couple on an allotment, garage roof, balcony or roof in a city" without any consideration for their neighbours.

Why is it people seem to forget *HONEYBEES DO STING!*

It is at this point they also realise how irresponsible and selfish they are being in their demands of how, sometimes, beekeeping is portrayed to them in the media in being the *"in thing"* to do.

Being eco friendly and doing their bit in saving the Honeybee.

This is irresponsible on both parties, unless beehives can be kept and looked after in the right location, don't even go there or think about it!

It also makes them realise how heavy the hive or supers can get when full of honey. I also make people realise they have a duty to the neighbours as they may like to admire a WBC beehive in the garden but their neighbours may not especially having to keep their windows closed on a hot day through fear of being stung while you are inspecting the bees because you have to do it on that particular day, being your day off, due the possibility of finding queen cells, due to family, work commitments or due to your neglect, collecting your swarm from next door.

If anybody gets stung and they know you have a beehive whether you have bees in it or not they will blame you.

The reasoning for doing this initiation is to show people how large a colony of bees can get, now if they show fear then beekeeping is not for them and regardless how much they want to help the environment then their efforts might be spent elsewhere in another area but definitely not beekeeping as these people would become a liability to all around with them losing interest and neglecting the bees and one of us beekeepers having to tidy up behind them.

What you should see and feel when confronted with a full size colony is an excitable adrenalin a sort of "wow" factor but being very calm at the same time as the bees can sense any tremor in your hands or change of temperature when the hive is opened up for inspection so *YOU* must be in control at all times. If you crush the bees you will be releasing the bees sting venom which will start them attacking you and everyone around so everything you do must be done in the bees interest, with very gentle movements and complete respect for the bee.

There are some beekeepers, which, just by watching them you know they are true beekeepers, and can handle a colony with little disturbance and with no smoke creating less reaction from the Honeybees.

There is an art in handling the bees; if you haven't got that passion for this kind of artistry then do not put the bees through any unnecessary stress. Your family, friends and neighbours will thank you for it. Do it for all the right reasons, don't do it because it is eco friendly by putting a beehive on a roof top or balcony in a city, even putting beehives on an allotment unless there is at least 100 yards between your hives and any other person. Find out how to deal with the bees properly respecting other peoples views, privacy and space after all they may not agree with your enthusiasm when being stung.

Preferably put your beehives out of sight due to honey being in very short supply it makes the beehives very tempting, if they can be seen they will be taken by a small minority of unscrupulous beekeepers who are quite prepared to take them, you can't get honey without bees so protect your investments.

It has happened to me, 10 beautifully refurbished WBC beehives freshly painted, put onto a Garden Centre well away from the public but stupidly in view.

They were there for 3 years before two beekeepers decided they would like them, due to the beekeepers protective bee suit no one was suspicious of there actions in stealing them, unfortunately it happens, £3000 pounds worth of equipment and lost honey today and in the future.

It does leave a very sour tastes in your mouth because you know it was beekeepers. Use my experiences to safeguard your own Honeybees and equipment, don't put beehives in view!

If you're not prepared to do any of the recommendations, then don't keep bees.

Due to the current situation beekeepers are in at present, finding an answer and solution to why honeybees are dying worldwide along with queen rearing is now more important and critical than ever in maintaining our industry to save the honeybee and produce good quality native queens too deal with our changing environment and climate.

Whether you go down the route of politically correct global warming, and everybody has a viewpoint, it does not matter what you call it or whom you want to blame. It still comes down to a word that we have always used and that is,

"Nature"

It will always sort itself out globally which Mother Earth has done for millions of years. You must consider the time span that human life has been here; possibly less than a single dot on this page compared to the size of this book if that was the time of our planet. Mother Earth is more in control than you think.

However, we can and must influence what strains and sub-strains of honeybee are left. As a starting point they must develop here with the native honeybee. Everything on this planet has evolved for a reason to be where it is and why its there. The climate over millions of years has dictated that.

Many of you that have successfully mastered the art of queen rearing, some have been in a situation to bring in other species to try to improve on what we already have, whether that is good or bad you should be proud of sustaining good quality queens year after year. Had the same effort gone into tracing a native strain rather than an exotic super strain we just might have been better off today than where we are now.

I feel that every beekeeper should rear their own queens, what you breed today will be tomorrows future queens in sustaining beekeeping worldwide. It is only then you can call yourself a beekeeper rather than a keeper of bees.

Everybody has his or her system of beekeeping and queen rearing. Each system works because the bees will always put it right even if it is wrong. When the beekeeper has gone the bees will put it how they want it. All you can do is applaud those individuals for thinking along the lines of what is best for them.

However, sometimes, they may not have had the time to experiment with different systems to find out what is best for the honeybee and therefore causing unnecessary stress within the colony.

After 30 years of beekeeping my Queen Rearing System has evolved as explained in Chapter Three. It has been developed from varying experiments and has been in production use of rearing queens for the past 20-ish years. Primarily written to appear in the BIBBA Publication, now a reduced re-edited version by *Roger Patterson* is on the website of the same.

Within the experiments I have tried to find a natural answer to the Varroa mite and *"with reading what I see"* I have discovered, developed and put in use, *Electromagnetic Geopathic Stress Lines, a Mini Nucleus System and a Debris Floor* that have all worked very successfully *without using chemicals for 18 years*. More on those three items in Chapters One, Two, Four and Five.

The main purpose of any system I produce is versatility in order to have additional use so that you always double its value in purpose, and its not waiting around for the best part of the year unused. All the equipment ie; beehives, nuclei, supers and mini nucs I have made is with tools that can be purchased in most DIY stores which are of basic construction and nominal skill required.

The reasoning behind my ideas and passion came from the very first hive that I purchased. This is a story in itself. I bought 2 National hives, one with bees in and some empty equipment. The contents had not been inspected by myself but I was assured that all was well inside by the previous owner. First mistake! Always check the contents.

However I was still not sure as to why the bees were aggressive when collected. At that time I thought this was normal. How wrong could I be.

After a couple of weeks in their new home I ventured in, only to find the hive was queenless, with no brood at all, hence explaining why they were aggressive. *A great start can't trust anyone, not even beekeepers!*

It was the middle of August and every avenue that I went down I had the same answer. I could buy a complete hive or nucleus with bees and queen but not just the queen, even though all I wanted was the queen. Eventually after a recommendation, I approached a beekeeper 70 miles away. It seemed a long way to go in order to obtain a queen but it was either go or give up. At the time I knew nothing of all the problems of introducing a new queen, or the possibility of a laying worker. In she went and I was lucky, I had a great queen that lasted 5 years. The best £6.00 I ever spent! *My faith and trust in beekeepers was back but what a learning curve.*

After all this, I vowed that I would never be without spare queens. The following year I started trying different queen rearing systems; Snelgrove, Demaree, Webmore and any book that mentioned queen rearing. Each year, a different way was used, varying and combining where necessary. Trying to improve on the different queen rearing systems available. Failing miserably some years and other years having fantastic results. Over the years the experiments and knowledge gained were improving. This resulted in the system becoming what it is today. The experiments over the years, did increase my spare queens with each queen becoming another production colony increasing my collection up to 300 beehives by 2004, *which may have led to my divorce.*

My expertise in making beehives got better, the woodworking tools increased resulting in a specialised workshop at the bottom of the garden. *My ex* preferred the noise being as far away as possible, hinting occasionally but really joking, so I thought, about moving a bed in there as well, due to the time spent being in the workshop.

Why is it that when the honey money is available their attitude changes, I will never understand…. mmmmmmmmmmmmm!

No doubt I was spending too much time with the bees and making more equipment in winter months always planning and thinking ahead to next seasons requirements, repairs and experiments, however much I made it never seemed enough.

"Unfortunately, my wife had enough and left in 2004 with inevitable divorce. It set me back, financially, plans for a new build house with a separate Honey Extraction and Woodworking Workshop facilities became a financial disaster. More important, emotionally, much more than I expected.

Work with the bees came to an abrupt halt, they were left to survive on their own for 4 years as; *My world was shattered.* Never realising the remarks, comments for my passion for beekeeping would come to this. I never thought she would leave, my life was devastated, unsure which way to turn, trying to find solace in other ways".

It is only in the last 12 months (2008) that I have got my life in order in getting back into beekeeping with my quest to finding an answer for the Varroa mite, everything you read has a reason and is all relevant for its actions in finding the end result! Anyway! *Why not "me" finding it? Somebody has to find an answer! Why does it have to be chemical or scientific?*

What of the bees you ask? Well yes, a lot did die out; but the ones that survived will be great breeding stock for the coming seasons.

BUT after further research and questioning. Why? Why? Why? The hives that survived, Why? Why were they so different? Why, with no management for 4 years did they come through? Was it my excellent breeding of native queens? Why was there very little or no Varroa? I didn't know then! These hives gave me the confirmation, answer and solution to finding out how;

"Honeybees can deal with the Varroa Mite without using chemicals"

This is explained in Chapter One *"Electromagnetic Geopathic Stress Curtain Lines".*

Yes, I can hear you say, *"every cloud has a silver lining".*…. get divorced then find the holy grail of beekeeping!!! Ok lol!

Beekeeping is like a jigsaw puzzle, if there is one piece missing the whole thing fails, as the picture is incomplete. Every piece counts to keep the structure together; beekeeping and queen rearing is just the same.

This is my puzzle, collecting together all the observations gained from 30 years of beekeeping experience and piecing it together!

With Queen Rearing: If you take away the queen, split the colony, take away the foraging network or the brood for any system you wish to use then you will be causing this queen rearing puzzle unnecessary stress. This could lead to failure, more often than not creating sub standard queen cells lacking in royal jelly and size. We must all agree that we will never domesticate the honeybee regardless of which type of box we put them in. *We have to read what they want and need for their micro-environment, wild existence, while in our care.*

In years gone by I have been called out to remove bees from varying places and why, oh why, do they always seem to pick the most awkward places? Choosing chimneys, old trees, and a cavity in the family house or stately home. All of these places have unlimited space below so debris is not a problem to them. But why choose this anyway? What drew them to these places?

However, there is a reason; *Electromagnetic Geopathic Stress Curtain Lines* I am aware this idea may cause some controversy because you can't see it, touch it, sense it or feel it but it is there and travels skywards to approximately 30,000ft.

My own research and testing in this area is still ongoing but it is conclusive, I just need to bring it into the 21st century as you will read later using digital equipment.

All the largest nests I have removed are from chimneys. Normally you would say light a fire to create smoke but if the nest is established with brood then no way will the bees leave. If removal is necessary and insisted upon, by the homeowner, due to them not wanting to see the bees destroyed and you are able to assist them. To achieve this it will be necessary to take down one side of the chimney to expose the honeycomb. That is of course after the homeowner has arranged for scaffolding that complies with health and safety regulations. Then, by placing on top of the chimney, a national hive brood box full of foundation with one frame of unsealed brood in the centre of the brood box and with a little help of smoke, the bees can be coaxed up from the wild comb into the brood box. On each occasion I have seen the queen walk in to the brood box.

The wild comb can now be taken out, honey and all. It is a very messy job, have plenty of spare buckets. On this occasion the nest was 7 feet in length on 6 combs with approximately 80lbs of honey that was eventually extracted.

What would have happened had you had lit a fire? Possibly burnt their house down, as beeswax is very combustible, so the right decision was taken.

The bees are now in their new home inside a national hive brood box on top of the chimney. They are left to clean up all the residue honey which can take them up to a week depending on the weather. When completed, the hive is then removed to a new site, more than 3 miles away, so they don't fly back.

The chimney can now be completely dismantled, then rebuilt with new bricks; the old bricks that came off must not be used as these are covered on the inside with *propolis* that will attract next years swarm. You will always be told that bees have been there for the past 50 years or longer, with the homeowner not realising that the bees had died out at varying times and have then been replaced by a new swarm attracted by the *propolis* and an *Electromagnetic Geopathic Stress Line*. With the homeowner assuming it's the same bees. You can remove the bees, wax, pollen and honey but you cannot remove the *propolis* or the *Electromagnetic Geopathic Stress Line*.

Both of these natural phenomenons are what attract the swarm. Once the sun has warmed the chimney the pungent smell of *propolis* and the *Electromagnetic Geopathic Stress Curtain Line* is to irresistible for that ever increasing colony that is getting ready to swarm in a nearby beehive that is unmanaged and in the wrong place. If proper ventilation chimney screening is used, verified by housing regulations then so be it, use this to seal the chimney but still allowing it to breath, however I have found new bricks removes any problem with *propolis,* and screening stops access. You can never get rid of the *Electormagnetic Geopathic Stress Curtain Line*. At least this action will eliminate future problems.

Most chimneys are not required today, due to central heating, and end being taken down anyway. *If you are unable to do it safely with the correct scaffolding then don't do it.*

"I will always remember being shown a photo of an old beekeeper balancing on the ridge of the roof to get to the chimney where the swarm had gone in. Whether he was successful or not, I don't know, but more fool him for doing it!"

As I told him! *Don't do it, if it's not safe!*

It appears, as we all know, that given the choice, honeybees would prefer vertical narrow empty spaces with unlimited depth. Just enough space to build side by side 5 to 6 combs but going up to 6 feet in depth or more with comb, with all debris falling well out of the way. If this were in a tree trunk with varying sizes of passageways, the bees would still interlock the colony with comb so that the queen or queens could cover all passages. Due to the deformed tree shape two queens are not uncommon in this situation, again with all debris falling out of harms way.

It is due to these observations my Queen Rearing System, Debris Floor and confirmation of Electromagnetic Geopathic Stress Curtain Lines evolved to what it is today, in *reading what I saw*.

I have over the years taken note of the equipment of previous generations that one inherits and wondered at the ideas and ingenuity that went into it. Nothing is ever discarded; it is either modified, converted or the idea carried on until it falls apart in testing or literally falls apart. After all, previous generations has put in the same effort, time and thought as the current generation and how many times does an old idea that has been lying dormant, become a new idea.

Many ideas sadly going to the grave. What is it with beekeepers? Why is it such a secretive mysterious craft? Such a waste of a lifetime's work and passion but unfortunately we have all heard of it. Old wives tales, folklore are so easily ignored or forgotten in this day and age perhaps because it was not seen on television or it is even laughed at…(mmmmmm! that just might sound like *Electromagnetic Geopathic Stress Curtain Lines*). Yet previous generations were very much aware of and used these lines to their advantage, being closer to nature than we will ever be.

Think about it!

Where did my ideas come from, you ask?

With beekeeping I have always questioned *"why?"* with each answer becoming another question and another piece of the puzzle, observing what I saw and questioning all the time, so the thought, logic and realisation behind my discoveries, might now become apparent. I hope you will be captivated with the following;

Chapters
Of
Amazement!

Please Enjoy.

An HOLISTIC *Way In Saving The* "Honeybee"

With

A Natural Phenomenon!

"Electromagnetic Geopathic
Stress Curtain Lines"

Plus

2009 Update

"2009 Update"

From further observation, research and work done this year, 2009, I can confirm without doubt that *The Varroa Mite* can either be controlled or eradicated by the honeybees, if your beehive is placed above an *Electromagnetic Geopathic Stress Curtain Line*.

Three things must be remembered with my discovery;

I. The earth does vibrate at a constant rate of 7.83htz unless disturbed.

II. Honeybees use vibration to control their micro-environment ie;

✓ *To regulate temperatures to an accurate level that will keep the brood warm, maintaining their existence and own microclimate.*

✓ *As a defence mechanism to ward off predators with sound produced by wing movement and vibration.*

✓ *Used as communication.*

III. Honeybee beehive numbers have been declining ever since mankind put them into logs, boxes, skeps or beehives. However, as a guide, from records we have, the numbers are quite clearly stated in the National Audit Office Report dated September 2008. This does tie in to their ongoing decline since man put honeybees into a man made product in modern times to harvest honey, as seen below, *quote,*

✓ *1900 approx 1 million beehives (pre Varroa)*

✓ *1950 approx 400,000 beehives (pre Varroa)*

✓ *2008 approx 274,000 beehives (unquote)*

✓ *2009 declining*

- ✓ *Honeybees have been on this planet for millions of years without any problem.*
- ✓ *Mankind found bees producing Honey, then put bees into a box or skep to domesticate and to harvest, Honey!*
- ✓ *That was the start of the decline now being what it is today after only a few hundred years of mankind's interference! Many years before the Varroa mite or any other modern manmade product turned up!*

Does that make sense!

These statistics and more do clarify my hypothesis!

I hope you will agree that I should receive recognition for;

Discovering the Answer & Solution

To

"An HOLISTIC Way In Saving the Honeybee"

With

A Natural Phenomenon!

The Explanation

"An HOLISTIC Way In Saving the Honeybee"

With

A Natural Phenomenon!
"Electromagnetic Geopathic Stress Lines"

This is the contentious bit because I am not sure how to explain what I have discovered, it all happened by accident with linking circumstances of events, yes, even including *my ex* leaving, plus with what happened over the last 18 years of no chemical treatment, experimentation and research, especially as it is quite unbelievable and really does work. It is far easier to show you than to explain. Anyway here goes…

It started when I read a couple of books, the first some years ago, one was about "Listen to the Bees" by Rex Boys about a BBC sound engineer (Eddie Woods 1964) who used technology of the day to record sounds of his bees, measured in hertz, to determine if there was any different sounds created at specific times like; swarming etc. It was called an Apidictor. I too tried the same idea, confirming the variations of 190htz-250htz, but used more modern instrumentation that set the correct note on a guitar and using a mini microphone inside the brood area plugged into this gizmo. Yes the results where quite valid with it working quite effectively so the idea of the experiment stuck with me for a very long time, as I felt is was the way forward, somehow, but how?

The other book, some years later, was relating to cancer research that became more pieces of my ever increasing evolving puzzle. These where stored in my memory banks, wondering if either would be of any use in the future. Not realising the link to the combination of the two was staring me in the face, as both had related to the same vibration of 250htz.

The author of the Cancer book is, Mr Rolf Gordon of Dulwich Health and the title is "Are You Sleeping In A Safe Place?" priced at £6.95 plus postage (at today's price). It is with his permission that I am bringing his book to your attention, as my discovery will change beekeeping, as we know it, and put that little extra interest into these ancient crafts.

It is a new discovery for beekeepers to consider in what the bees want, but not a new idea in how to find it. However, it just may have got lost in time due to us becoming more civilised.

Mr Gordon made a reference to plants, trees, birds, animals and insects in his publication. As there was no name or book reference relating to insects I contacted Mr Gordon to ask for some clarification on his findings, why he made his references to insects and how it came about. However he was unable to help, although he did ask me to substantiate his theory with *"Honeybees"*, so that he may use my name as his reference within future editions of his book.

This I believe I have now proved conclusively in what you will read.

Further conversations with Rolf left me in total amazement as what he had to say concerning the content of his book. To his reasoning and his personal interest, and the tragedy behind his passion for research in this area of Cancer.

I think you would need to read his book, now in its 7th edition, to totally understand where I am coming from. However you do have to have a completely open mind as it is very thought provoking.

Some of you will think that it is a load of rubbish whereas others will try it even if it's to prove it wrong! Some of you will be totally amazed in finding that it works and use it as part of your basic beekeeping in keeping the Varroa mite and possibly anything else at bay.

There is nothing to lose by trying it.

There are situations within this planet and even concerning the human brain which scientist and doctors are still learning about when certain things happen that cannot be proved or disproved by science. However you wonder at it because it works. This is one of those moments that will stretch your imagination.

We know that given the perfect opportunity in the wild state, honeybees will build their comb directed to magnetic north/south when space permits, why?

We don't know, but they do!

Has any scientific work been done on this observation, possibly not! So why do we expect the bees to do otherwise when we put the combs the other way around i.e. west/east when we site a beehive?

We expect honeybees to conform to what we want and do not take into account what they prefer. This may cause them unnecessary stress as they are positioned against their natural building direction.

Are we bothered or do we really care?

I have always wondered *"why?"* and with modern day technology a query or clarification can usually be answered with the written word, television or the Internet on your personal computer.

The access to knowledge that we have is vast, beyond our wildest imagination. Unfortunately much of the information that our elders had learnt is being forgotten. For example, old wives tales, folklore, sayings and skills are being lost, generation-by-generation.

Our fore-fathers were closer to nature than us, as *Nature was their television*, and had therefore learnt to understand it, but what remains of this knowledge is usually disregarded in this present day, ignored or laughed at, as not scientifically proven.

Where am I going with all this?

Possibly to drop it in gently hoping that you won't think there is an ambulance with attendants in white coats waiting to take me away…

Please bear with me…

There are things in life that you wish you had thought of first, because they were so simple, being in front of you all the time, and you not seeing it, you could not see the wood for the trees scenario. Does that sound familiar?

There must be times in your life where for a brief moment you have wondered about nature asking why? or how? but then carried on and brushed it aside thinking no it is not important.

Have you ever wondered or asked yourself the following.

- ✓ *Why do honeybees settle to make their home in a particular place when swarming?*
- ✓ *Why do honeybees sometimes choose the smallest of cavity knowing it's not big enough?*
- ✓ *Why ant's nests always seem to be in a particular place every year, even though they may be further away from their food source?*
- ✓ *Why sheep graze all over a field during the day, but sleep in only one part of that field at night?*
- ✓ *Why Oak trees are always in a line even though the acorns fall all around but will only germinate in one place?*
- ✓ *Why are Oak trees more likely to be hit by lightning than any other tree?*
- ✓ *Why does Elderberry also grow in a straight line?*
- ✓ *Why a single tree on its own is out of shape and distorted?*
- ✓ *Why fruit trees will blossom but not fruit?*
- ✓ *Why certain plants thrive but others do not?*
- ✓ *Why cats prefer to sleep in one place?*
- ✓ *Why the UK has the highest number of cancer patients?*
- ✓ *Or even cot deaths!*
- ✓ *Birds, travelling thousands of miles when migrating?*

The answer will probably be NO!

Why would you want to know anyway? You are running around in your very fast lifestyle. After all it is only nature and you can't do anything about it or can you? Can you use this information to your advantage?

Yes! Because all the above points are affected by and indicators of;

"Electromagnetic Geopathic Stress Curtain Lines"

A Natural phenomenon

We don't look at nature anymore; we lead a very fast lifestyle and go around in our cars with our eyes blinkered. Not really caring or being that bothered about what is happening around us even though are fore-fathers knew of this phenomenon and used it in various ways to their advantage.

I am not talking about being an eco-warrior, that is something else, I'm talking about reading nature and what it has to offer us, now, today.

Our forefathers did exactly that. They would look, listen and learn these skills concerning the natural world and teach these skills of nature to their children, then pass it on down from one generation to the next. They knew that this knowledge was tried and tested and help to sustain *survival* for the next generation. Here is an example;

"The Vikings and Romans would use sheep for food, but also they were used as a tool to find where to put new settlements. They would only build where the sheep slept, because the sheep would try to get away from any unnatural earth force that may affect them while sleeping".

Why is it today we build houses anywhere, without checking what is beneath? In many countries this is standard practice in checking before they build. This leaves problems of continuing unexplained illnesses that will lie ahead for another generation to solve not realising the answer is beneath them…

Due to sophisticated and civilised living, these natural abilities have been stifled and squashed and are no longer part of daily life. Probably laughed at in this modern day, not understanding how nature was formed and how long it took to be so perfect as it is today.

So what is it that has been used for thousands of years yet termed as witchcraft? Condemned by the Church 500 years ago proclaiming that *it is work of the devil!* As I said earlier Mother Earth does vibrate, NASA know of this but if you were to ask a 100 people down any High St or Shopping Mall the answer would be the same, because it *can't* be felt the assumption is that Earth does *NOT* vibrate unless there is an earthquake or volcano!

The earth does vibrate at a very low constant measurement which your brain counteracts so you don't feel it. This measurement is measured in hertz and the rate is 7.83htz.

Remember the human race has had a very long time to adjust to this vibration after all it has been there since time began.

It is this measurement that Rolf Gordon has researched with cancer sufferers. I will leave it for him to explain that in his book. His expertise in this field of cancer research and general health, which you will find both interesting with compelling results. It is worth a read. "Are you sleeping in a safe place"?

- *Have you ever wondered why NASA had problems of illness with astronauts in their early flights?*

No, course not as they did not publish it then as it was still secret. It is because of this low vibration 7.83htz that NASA had early problems with their first astronauts going into space.

There is no vibration in outer space, in their craft or suites so the astronauts became ill, (their brain was still counteracting the vibration), and could only live in space for a short time. In the subsequent flights the low vibration of 7.83htz was installed within the Craft, Suites and Skylab with the astronauts able to stay in outer space as long as humanly possible. *"I did confirm this when on a visit to NASA Houston Texas".*

"The earths normal low vibration of 7.83htz gets interrupted by hollow chambers of running water/fluid creating friction allowing oscillation to resonate to become an **Electromagnetic Wave Vibration** *which will increase upwards to 250htz or more which is 30 times greater than surface vibration of 7.83 hertz."*

Now you cannot get rid of them because they are in the make up of the planet and they are everywhere; millions of them, throughout the whole planet. Like rivers, below ground, some are very small (4inches wide) and some quite large (4feet wide) travelling straight for miles. These rivers are at a depth of 200 to 300 feet or more crossing over each other at varying depths zig zagging there way across the planet. Ultimately interrupting the natural earth vibration and continuing to the surface and skywards, up to 30,00ft, as a *curtain line* of *Electromagnetic Wave Vibration*.

These *curtain lines* have been there for millions of years, and therefore plants, trees, insects, birds, animals and human life has evolved with it and can sense these *curtain lines*. Life has been created over millions of years, human, plants, birds, fish etc, are either drawn too these *lines* where they will thrive or repelled by the vibrations doing them harm so move away as they will not survive. This is the evolution of *"Nature"* that has evolved over millions of years creating what we have today.

Our own babies, if above one of these *curtain lines*, can sense it while asleep in their cots and will try to move away from it, as we grow older we lose this sense. (Rolf Gordon)

If only the medical profession were not so blinkered or hell-bent on giving you a chemical cure when there just maybe a natural remedy that will do the same or even prevent it happening in the first place; it does make you think?

Honey for example has been used in medical matters for thousands of years but do doctors prescribe honey????

My own General Practitioner Doctor in the last few years has kept my honey in the surgery as he was getting fed up with people going to see him for hay fever and yes honey does work in this area if you take the right dose at the right time therefore saving the National Health Service millions. If all GP Doctors did this, what a boost to beekeepers and Local Honey.

What I am going to tell you about later could save the NHS a lot more, but then it is so frustrating as a beekeeper, we know of the medical properties of honey which can do the same as a billion dollar chemical pill. Sorry, got on to my soap box, we carry on… another book possibly.

Oak trees, Elderberry, Mistletoe, Mushrooms, Asparagus, Cats, Ants, Wasps, Beetles, Birds, Fish and much more are all drawn to these *Electromagnetic Geopathic Stress Curtain Lines,* and especially,

"Honeybees"
They love them!

Attracted to G S lines like **magnets,** *in fact*
Electromagnetic Curtain Lines!

Now wouldn't it be strange if the Varroa mite hated the higher vibration of 250htz? Or the bees were able to use this higher curtain line vibration to control or eradicate the Varroa mite? No, we could not be that lucky.... (Or could we? Read on, and you will be pleasantly surprised) *A natural phenomenon able to solve a major infestation of the Varroa mite, and whatever else that is infesting honeybees worldwide for the cost of;*

NOTHING!

- Does stinging nettle and dock leaf come to mind?

- The answer is there but we cannot see it. Possibly the part of the brain that senses it can only pass on this information involuntary, through the nervous system. However we have lost the use and need for this through the passage of time.

- Lets just say you heard it here first. My tests and research are still ongoing in using modern technology rather than using the ancient method which is very accurate and is still to be mentioned.

- We owe it to the bees and ourselves to find another way, a completely different way without chemicals.

- Mankind has put honeybees in this precarious way.

- Mankind has moved Honeybees from their natural habitat so man could harvest honey.

- No one has considered, including our greatest scholars and scientist, until now, *by me*, in asking;

"What do Honeybees really need to survive?"

So given the choice... Is this where they would prefer to be?

"Placed in their beehive above an Electromagnetic Geopathic Stress Line Curtain which is 30 times higher vibration 250htz, than normal earth vibration (7.83htz), where honeybees are attracted to every time they swarm (proved). Woods recorded the same vibration 250htz in the bee nest. Does this mean they have less work to do? Yes! The higher vibration is used to their advantage being able to concentrate and manage their micro-environment better. Which then results in a more harmonious, healthier beehive, without the Varroa mite (proved), back to their natural state of survival and a honey yield that is increased 2 or 3 fold (proved)"

Now, I think, I may have your attention!

We must look elsewhere other than chemicals!

A prime case, to the attraction to *Electromagnetic Geopathic Stress Curtain Lines* happened in my own garden. Spare equipment had been brought back for refurbishment; a swarm appeared a few days later, and settled into one of the spare hives. It was only when showing a friend how to dowse that it became apparent the swarm had a choice of many empty hives, however they chose the one that was over an *Electromagnetic Geopathic Stress Curtain Line.* *"Thereafter every location was checked where swarms had settled, each was above an Electromagnetic Geopathic Stress Line. Thus confirming, yet again, this is what they are attracted to and need for survival"*

When swarms are collected how many times have you noticed them fly away into the distant horizon after you have carefully positioned them onto its new site? It is because you are putting them in the wrong place away from an *Electromagnetic Geopathic Stress Line!* We have always in the past scratched our heads and put that down to beekeeping. Well, now you know why they flew! They knew where they wanted to be!

I can confirm that each of my hives that survived, without any inspections for 4 years, since my ex left, has NO Varroa mites and honey to spare, each one was above an Electromagnetic Geopathic Stress Curtain Line. How much confirmation do you need, this is what they want, it works.

I can categorically confirm that each of you has one or a few hives that out perform others with *You* not sure as to why putting it down to Queen mating or strain resistant to Varroa? *(That's what I thought! I am sure if you check those hives they will be above an Electromagnetic curtain line).*

Their stress levels are at a minimum due to not working so hard (30 times less) where they are more harmonious with nature, and able to deal with the Varroa mite and all that comes with it.

Electromagnetic Geopathic Stress Lines; you can't see them, you can't feel them, you can't touch them, you can't sense them and you do not know where to look for them but they are there! My own house and garden has *8 Electromagnetic Geopathic Stress Lines* proving they are not miles apart but within a very small area crossing over each other almost like a very large cobweb and you not knowing they are there. I would love to have an aerial view after plotting where they are just to see if there is a shape to these *curtain lines.*

How Do You Find Electromagnetic Geopathic Stress Lines?

This is an old art, skill, or if you want to call it what the church did, *witchcraft,* that is very much in use today and which anybody can do. Most people would say it was used in olden times just to find water but it is still used today in big business.

They use this method to find minerals, pipes, oil, water, tunnels, mines, diamonds, gas, if fact, you name it you can find it by this method and it is called;

Dowsing!

using

Divining Rods!

Yes! I know what your thinking!! You were expecting some thing out of this world something completely new, well it is new to beekeepers. Divining has the answer to what it can find, an old idea lost in time, however don't stop reading now as the best bit is yet to come.

I have already mentioned that it is not new; just forgotten by time in our daily lives due to being more sophisticated and civilised because we do not need to use it anymore.

We do not need to find water because it is piped to us. Everything that we want is readily available so we do not need to look for it. We just turn on the tap or go to the local supermarket.

Honeybees are attracted to and benefit from being above these lines but do we care enough to find them? I certainly hope so. If the bees are at one with nature then that must mean they are more productive and healthier and keep everything that should not be in the beehive to a minimum or eradication, yes, even the Varroa mite. Remember this is where and what they would have selected if given the choice to stay when they swarm.

Divining Rods, How To Make Them

All you need are two bic pens (optional) and two wire coat hangers. Trust me as this really does work.

Take out the pen ink inserts and discard; you need the outer as a sleeve and holder for the rods.

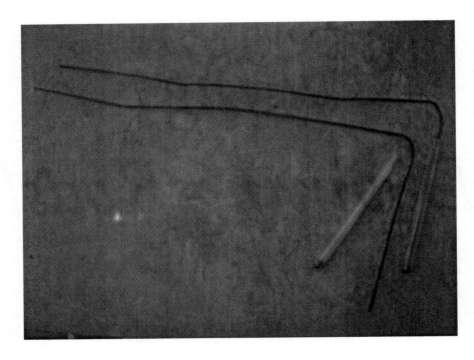

Cut your coat hanger so you have an L shape and bend accordingly, so that you have one length of about 15 inches with the other being shorter. Place the shorter end into the plastic holders (optional), as seen in the photo. What could be easier to make than that. You can use thicker wire if available.

That is it, you now have two divining rods that really do work and anyone can do it. Yes, anyone, and its cost you nothing. How does it work? Nobody really knows but it does. It has been used for thousands of years, successfully.

Science can neither prove nor disprove how it works. It is a combination of part of the brain that we do not understand along with energy of the nervous system within us.

Try it!….. No harm can be done and once you get the idea just think of the return in healthier bees and bigger honey crop and the fact that it is free. You have nothing to lose!

The more you practice the better you become.

How To Use

The wire rods should be free to move in your penholders (optional), with one holder in each hand, making sure your thumbs are not by the wire, hold them out in front of you parallel to each other. I hold mine slightly downwards at a very slight angle so there is far less likelihood of a false reading, not that I ever do.

Now think, think and really think… *Electromagnetic Geopathic Stress Curtain Lines.* If it is windy outside try it indoors so the neighbours do not see you. It might save your blushes. When you slowly walk over an *Electromagnetic Geopathic Stress Curtain Line* the wires will cross over each other and show direction of the line. Yes scary the first time.

When it happened to me I screamed *"!~:<* well, something like that. I was told, by Rolf, over the phone where that line would be. He was completely accurate, that is why it was scary. The more I checked the easier and better I became.

It may boost your confidence if you dowse knowing that a line is there. For example; find a line of Oak trees or Elderberry and walk in between two trees. I can guarantee a reaction from the wire rods that will help give you a *"feel"* for the rods and make it easier when checking virgin land. Try it! Dowsing can be used to find anything, you just have to think of nothing else except what you want to find. It can't be easier.

The Egyptian Queen, Cleopatra, had two dowsers with her at all times to find precious stones, silver, gold and much more with great success. Ancient civilisations thousands of years ago, Druids, Persians, Chinese, Hebrews, Egyptians as shown in the hieroglyphics, Hindus, Greeks, Romans and American Indians all knew of dowsing, Moses was an expert dowser as stated in the Bible. Were these people stupid or where they using natural earth resources for survival.

They knew but may or may not have understood the scientific value that divining could be trusted and was available to them from nature using twigs from a certain tree as their rods, think about it!

If you want to explore further with this technique then information can be found from The British Society of Dowsers where you can buy heavier rods that will be better for use outside just in case there is a slight breeze. However it is quite easy to make your own.

✓ *Well, is it worth you trying?*

✓ *Because it really does work!!!*

✓ *What's it going to cost you? Nothing!!!*

✓ *A word that most beekeepers like!!!*

✓ *Do you really want to be Varroa free?*

✓ *Using No Chemicals, Ever!!!*

✓ *With an increased honey crop!!!*

Here are 2 simple case studies for you to do.

First;

In spring, find an Electromagnetic Geopathic Stress Curtain Line and place a colony that you know is riddled with the Varroa mite above the line. Face the edge of the frames magnetic north/south. Leave for six to eight weeks with normal inspections. During this time you will see many deformed bees moving away from the hive. After eight weeks you will see very little or no Varroa mites.

The bees will look healthier and more pro-active producing more honey. If it has an old queen she will be superseded not long after being Varroa free rather than swarm.

If left in its old site this hive would have perished. I hope, you have plenty of spare supers, you'll need them!

Here is the second;

Wait for your first swarm, then test with your wire divining rods where the bees have settled. A guaranteed Electromagnetic Geopathic Stress Curtain Line!

Benefits

*   Varroa controlled or eradicated completely.

*   A higher crop return… 2 to 3 times greater.

*   Your bees are under less stress.

*   Possibly easier to handle… mine are.

*   Bees wanting to be above a higher earth vibration to compliment their own vibration.

*   A healthier hive due to greater numbers of brood not effected by the Varroa mite.

*   Are they trying to tell us something when they swarm? (That should create some debate).

*   Possibly other diseases being dealt with.

*   Saving the Honeybee from extinction.

*   Beekeeping as we knew it prior to the Varroa mite.

*   Honey in abundance.

So what if you "don't" do the test, could this be happening?

Is it possible their own bee wing vibration is trying to counteract the earth's low vibration of 7.83htz, due to being located, by you, in the wrong place. The bees then use too much of their time, effort and energy trying to equalise and replicate their natural electromagnetic vibration which is 30 times greater than earth's normal vibration, possibly 250htz; and now they can't cope. Pests and diseases are attracted and become overpowering, not being dealt with as they should be, with other duties like, brood rearing, general hygiene in clearing comb and intruders due the excessive pressure we have put them under. They are in the wrong place with the wrong vibration and polarity. Hence disorder, collapse, dying out or disappearance.

Does that make sense, if not read it again!

I ask you, is that a possibility? The answer is

So, we know, from what you have read;

- ✓ *Honeybees do use vibration 190–250htz to survive and communicate. (measured by Woods)*

- ✓ *Honeybees swarm from 7.83htz to and above a higher vibration line 190–250htz. (Verified by Author)*

- ✓ *Honeybees did not choose to live in a beehive.*

- ✓ *Honeybees can deal with and eradicate Varroa. (Author)*

- ✓ *Honeybees use Earths magnetic field.*

- ✓ *Honeybees do not need Chemicals.*

- ✓ *Honeybees use Earths Electromagnetic Geopathic Stress Curtain Lines for Navigation and SURVIVAL!*

"It is not too late to learn from the honeybee, there is still more that they can teach us! If I could put this into a bottle and sell it to you then you would be happy to pay a small fortune. However, it is in our nature as humans, that you will be sceptical and doubt it, because it is not proven by science and is free". Free from nature!

I am sure the following, I hope, just might get you thinking!

- ✓ *The native Honeybee in mankind's beehive.*

- ✓ *Placed above an Electromagnetic Geopathic Stress Curtain Line.*

- ✓ *A very pro-active Honeybee colony.*

- ✓ *Frames lengthwise facing magnetic north/south.*

- ✓ *Your management skills.*

- ✓ *With a Harding Hive Debris Floor.*

- ✓ *Very little or no Varroa mites.*

- ✓ *Other pests and diseases being possibly repelled by earth vibration so no use of chemicals, ever.*

- ✓ *2 to 3 times more honey yield.*

- ✓ *Possibly no feeding.*

- ✓ *Knowing that we "mankind" have saved the Honeybee from extinction.*

(Well, I did really, it was my idea!)

Would you really like to go back 30 years pre Varroa mites, enjoying your beekeeping as you knew it then, just having to worry about the weather and swarming. Who knows? It has got to be worth giving them all the help we can, rather than letting the *Honeybee* die out. After all, we have put them into this predicament!

- You can't see, feel, sense or touch them, but the Lines are there! Honeybees can sense them!

- I know it works and it's FREE!

- I told you it would be contentious!

- Is it worth you trying? You've nothing to lose!

- Is it worth helping your honeybees to survive?

- Why have mine survived with no chemicals?

- Why have I always had a greater yield?

- Why do we have to rely on multi billion dollar chemical companies to come up with the answer and then charge us the earth? All I am asking for is that the earth or the bees can supply the answer, all you need to do is find a

"Electromagnetic Geopathic Stress Curtain Bee Line".

- Does anybody know of any previous work on vibration or this subject of Geopathic Stress Lines?

- I do know of the work done by Rex Boys and his publication of "Listen to the Bees". A combination of bee wing vibration measured in hertz to tell the state of the hive by sound, trying to help with swarming prediction. I don't think he realised how close he was to giving us the answer for the Varroa mite, 50 years early.

- I am the first to combine Honeybee vibration to earth vibration of Electromagnetic Geopathic Stress Curtain Lines giving the answer and solution to save the Honeybee.

For those of you that are now saying;

"If what you say is correct then where are all the wild honeybees?"

It was not scientific but an assumption, on everybody's part, even mine, that the Varroa mite was to blame for the demise of feral colonies. However they are still there, in smaller numbers, but they are still out there. Have you ever been to look?

Yes, they have to deal with the same weather as our honeybees, for forage, mating flights, survival and allowing ours to swarm, even though they are declining.

Have you considered this?

Their natural habitat is getting smaller, politics has changed farming, and mankind is moving out of the cities into surrounding country areas. Not happy with fresh air and country living, mankind has over the last 50 years tried to be politically correct in changing the country to how they would like it rather than enjoying its beauty.

So animals, birds and insects like honeybees are losing their natural wild nesting sites and ending up in someone's chimney or wall cavity as described in the introduction.

I, like thousands of others have been called out to remove a swarm or nest of honeybees and this is a delight in itself to us. Once collected, their selected old site is now sealed up, not to be used ever again by honeybees. Now multiply this over the last 50 years by the majority of beekeepers. Also Council Pest control and Independent Pest control have given advice to the homeowner on how to get rid and stay rid of bees.

You can get a rough idea as to how many feral colonies, hundreds of thousands perhaps millions, have been removed from where they would like to be, above an *Electromagnetic Geopathic Stress Curtain Line*, only to be saved by mankind, whoopee!

BUT then they are put into a beehive where mankind decides to locate it, kills it in time with an overdose of Varroa mites, by putting it in the wrong place, so no swarms, no feral colonies hence increasing decline year, after year, after year, after year!

Honeybees have been reducing in numbers for over hundreds of years, check the National Audit Report 2008 (figures shown at the beginning of this chapter), its not just since the Varroa mite, all they have done is escalate the problem bringing it to the attention of the media and mainstream public.

In fact, ever since the first man put Honeybees into logs, boxes, skeps and eventually beehives to harvest honey, with no regard to where the bees wanted to be placed, but then how would we know? Nobody except me has done any research in this area as all the top scholars are looking for a chemical or bacterial solution funded by the chemical industry.

My discovery really does work, guaranteed!

Just perhaps, consider the possibility that I am right with my hypothesis, we do have a major reason for helping them, after all, remember if Einstein, alleged statement, was right and mankind does falter then it is all beekeepers past and present that have put Honeybees into this present situation.

We owe it to the Honeybees to put it right however I haven't got all the answers this is just one major answer and solution that will save the honeybee worldwide, lets hope it saves mankind as well.

Now that I have found the answer and solution, who else are we blaming rather than ourselves?

Pesticides, GM Crops, Varroa, Mobile Phones and Masts, CCD(Colony Collapse Disorder) Only in the USA, Global Warming, Climate Change plus whatever else as long as it is not us. Nobody wants the blame! But we are to blame!

Lets take each one using observation and a bit of logic, after all, that's all I am, a logical observer as a beekeeper!

Pesticides and GM Crops;

An almost impossibility as all flying insects would have died.
If it was an aerial spray then it is localised to that area, not everywhere. Both these would show as a local area problem.

Varroa;

The main executioner of the western Honeybee specifically in Europe, however, as you have read, the bees can and will deal with it when your beehive is placed as I suggested.

Mobile Phones and Masts;

Honeybees were dying way before a mobile was even on the drawing board. I am surrounded by masts but are my bees dying, *NO!* As an observer I would have said this is a more of an intermittent signal rather than a constant signal.

Global Warming & Climate Change

This is one thing I call Nature and always will. Whatever happens Honeybees will survive, well, they have for the last 200 million years during different climate and polarity changes, so it's not this one.

CCD(Colony Collapse Disorder) <u>only</u> in the USA.

So what has the USA got that we haven't as they have all the same problems with Honeybees as we do, with all that is mentioned before. If beekeepers in the United States followed my recommendations of *Electromagnetic Geopathic Stress Curtain Lines* would they still have a problem, possibly, Yes!

Why? The USA is the only continent that I know of, that has, a Nuclear Deterrent Frequency similar to that of a Honeybee. Being at a constant low level frequency transmitted every 200 miles or so across the country which interferes with the navigation of a Honeybee in returning back to the beehive. As I have indicated so far Honeybees rely on a natural phenomenon of Earth Magnetic Vibration using their own built in navigation.

Honeybees not placed above an *Electromagnetic Geopathic Stress Curtain Line* will then follow the same frequency vibration line (Nuclear Deterrent System) on their return to their hive, which they can't find, then drop, fatigued, hence why they are disappearing with no dead bodies near the hive. *This is a possibility!* However the only people that can verify this are the U S Government, with the above being the only common denominator the rest of the world has *not* got.

Will they? Can they? Would they? Is anyone being told to keep quiet? You will have to draw your own conclusions as it cannot be anything else.

Due to the shortage of Honeybees in the US, many Honeybees are being imported from other countries which show no sign of CCD, to boost the pollination contracts. However as soon as the imported Honeybees are in the US, disappearance occurs.

Beekeepers in the US then start blaming the country of origin rather than accepting a self inflicted problem. It does make you think and it does make sense as to why keeping Honeybees in the USA is different from the rest of the world.

"The demise of Honeybees is really quite simple, it is just owning up to the facts. Mankind has a lot to answer for, with our wrong doings in the past to the Honeybee, however we can rectify the problem quite easily with my hypothesis........

In the UK and rest of Europe, find and place all beehives above an Electromagnetic Geopathic Stress Curtain Line, stand back and watch the results.

In the USA, do the same as above and if disappearance is still happening then query your Governments Defence tactics. Allow me to prove my theory in the US using modern military equipment that is available and being used by the military".

From current information available;

Why is CCD not happening in Canada in great numbers except near the border of the US?

Why is every state on the perimeter of the US suffering from CCD?

It is nothing to do with disease or a cocktail of pesticides, ok the Varroa mite is not helping but they can be controlled without chemicals or bacteria. So, am I right with my hypothesis of CCD? Possibly! We may never know as it could disappear as easy as it appeared, only the US Government can answer that, leaving just the Varroa mite which is now controllable by the Honeybee using my hypothesis.

Oh dear no bees!
So, no swarms!
Or feral colonies!

In fact, no Varroa either!

Well Done Mankind!

How long has mankind got without the Honeybee?

Chapter Two

What The Future Holds
&
The Way Forward

What the Future Holds

If things don't change they will stay as they are or possibly get worse... so what are you waiting for?

There are always sceptics out there, who are sniggering; saying *"Yeah right"* and laughing at the whole prospect of something so simple being the answer. Stinging nettle and dock leaf scenario, comes to mind yet again...

Most things in life are simple!

These sceptics are most probably descendents of those who would not accept that the world is round; until of course they tried it. Why do you always have to have a scientific answer?

If it is left to the giant pharmaceutical companies to find an answer, who knows what they may come up with? One possibility that I have been told, at a recent meeting, quote...

"Is to put still further bacteria into the beehive that lays its eggs into the Varroa mite and when ready to come into the world it literally breaks through the Varroa`s mites body, just like in the film, Alien"

Unquote.

- Yes, it kills the Varroa mite but what of the bacteria that are left?

- What happens to them?

- What if they mutate?

- What side effects are there for the bees and for us?

- As we all know honeybees are good at storing pollen, propolis and nectar. Is there a possibility that the bacteria will find its way into the honey and is extracted and eaten?

- Does that bare thinking about?

- This is just one possibility!

- We must keep our hives clean of further contamination by these so-called experts!

I dread to think of what else is happening in laboratories around the world. They are playing God with our honeybees.

But then YOU would be happy because YOU have just spent a fortune in buying their product, assuming it to be safe, so it must be all right. Come on, open yours eyes. You are putting too much trust in these people! There has to be side effects, remember the first licensed products killed my Queens, which will only be found when you try these miracle cures. You and your bees end up being the guinea pigs. Will they be interested when it is too late or when it goes wrong. I don't think so!

There are millions of dollars being given, in funding, to research the possibility of finding an answer. This will then be sold to you. However most or all the top scientist, universities and professors are funded by the chemical industry. It is impossible for them to take on board my hypothesis because they would lose their funding. This in turn would cost the chemical industry millions of dollars in lost sales.

Am I starting to make myself clear?

What I have found is free and therefore very upsetting to large pharmaceutical companies that have invested millions. So far, without a chemical/bacteria answer.

Until Now! My Discovery!

I have not treated my colonies with chemicals, not even sugar, for 18 years as the *"approved"* chemicals, at that time, were killing my Queens!

What I have found cost you nothing!

What have you got to lose accept your bees!

The Holy Grail of Beekeeping has been found!

And I found it.

All you have to do is try out what I have discovered. Eliminate chemicals for good and save the honeybee from extinction.

The Way Forward

At the moment the only way to find the *Electromagnetic Geopathic Stress Curtain Bee Lines* is using divining rods. However the technology is there to find earth vibration, but the problem is that, it is very expensive.

I know what I have discovered does works. However I still need to research further into what the bees want in earth vibration. There are many questions that still need to be asked.

- *Is there an optimum hertz they would prefer ie, 190htz or 235htz.*

- *What level of hertz vibration do the bees require to dislodge or remove the Varroa mite from their bodies while grooming?*

- *What if Varroa doesn't like the higher vibration of a* Electromagnetic Geopathic Stress Curtain Bee Line *while incubating inside the cell?*

- *Would it be a different vibration for any other bacteria or minute alien for the bees to remove or deal with?*

- *Do the bees require an irregular vibration possibly to help with varying tasks within the honeycomb nest?*

- **This technology could also be used to identify bad sleeping areas eliminating poor health, possibly Cancer, saving lives and the National Health Service millions of pounds.**

Until the necessary funding is made available and the research is done we will never be sure...

Allow me to prove it!

As I said, the technology is already available via seismology, detecting earthquake vibration. Unfortunately the machine offered to me was £8000.00. This is too heavy and bulky to carry around and way out of the price range of most, if not all, beekeepers.

A hand held alternative is available but needs tweaking with research which requires funding. This would provide a hand held devise with a digital read out, recording and plug in facility to a PC or Laptop. The price would then be attractive to all beekeepers worldwide.

It would give you,

- *A lightweight hand held devise run on batteries with car recharge capabilities.*

- *A digital read out in hertz when you have found an* Electromagnetic Geopathic Stress Curtain Bee Line.

- *Guaranteed to reduce or eradicate the Varroa mite when above an* Electromagnetic Geopathic Stress Curtain Bee Line.

- *Plug in facility to your PC or laptop to down load your information. The information can be used for an array of different projects.*

- *Increases of 2 to 3 fold your normal honey crop.*

- *An easy way to detect lines beneath your house where you sleep helping to further Cancer research and other illnesses.*

If there is found to be a standard optimum hertz vibration then an alternative manmade product could be easily made to fit into the floor of all beehives creating the perfect optimum vibration throughout the colony. The hive could then be placed anywhere. Would that work? We do not know until funding for research is available. It must be worth a try!

FUNDING AND RESEARCH IS WHAT I NEED

A University,
That is prepared to help with my work.

Every beekeeper on this planet will want to own this hand held devise so the return would be huge.

Einstein allegedly stated "that if Honeybees die out, mankind would have four years to live"

Well, was he right or wrong? Do you really want to put it to the test? I think not. Do something about it as I am trying to do.

Once *Honeybees* are gone they are gone for good!

Well except mine!

React to the Book and Do

Something!

or

Buy my Bees when yours have died!

Do it now, Make some Rods!

Thank you for reading

what

the Future holds, but what are

YOU

Going to do

about it?

The future:

To save

"The Honeybee"

Is in your hands

Right now!

Chapter Three

The Harding Queen Rearing System using Two Queens.
(As seen on the Internet)

I have been very aware that beekeeping, is something completely different from normal everyday pastimes and so are the people that do it. I will try anything once. I have been part of groups; however I found that with my style of beekeeping and queen rearing it is better to work on my own.

Therefore a great deal of what you have and will read is very much my own efforts, which has evolved, as you know, into my puzzle of beekeeping, this being just another piece.

Some years ago, I explained to Albert Knight of BIBBA,

(Bee Improvement and Bee Breeders Association), at the BBKA (British BeeKeeper Association) Stoneleigh Convention, a system of queen rearing that I have found successful. It is because of Albert that I became a member of BIBBA many years ago and served briefly on the committee with BIBBA. After a few further phone calls to explain the system, Albert put my idea in for publication as,

"The John Harding Method of Queen Rearing"

It appeared in Bee Improvement magazine number 20, (page 21). Unfortunately he had not seen it in the flesh so he used his own interpretation of my explanation over the phone. It was put into practical use, by Albert, in 2006 and was successful each time it was used.

Hence my name was put to this system, permanently.

At the time I was not completely au fait with the use of word processing and I am still not proficient. I let the article stand, even though there were some things that required clarification. So, here is the renamed article with explanations.

This idea has come from years of research, heartache, and earache, balancing all sort of balls in the air and most of all, trial and error and by watching what the bees do.

However it is many ideas turned into one, and trying to be as close to the wild as possible.

The Harding Queen Rearing System Using Two Queens

Frame or Lid Method can be used, as shown.

Over the years, I have always made my own equipment based on National specification, so I had in abundance many 5 frame national nuclei. These were to be the basis of this project and would be multi functional. It is better for this system to make a purpose built stand that is stable and at a height that is agreeable to you. This will make manipulation far easier.

For the base nuclei, a 50mm tube or larger, connects all nuclei together. This will allow the bees to roam from one nucleus to the next giving end-to-end connection to all 3 nuclei.

Queen excluder should be used on the inside of the two outer nuclei to keep the queens from entering into the central nucleus. Each base nucleus has a small central entrance to the front and sides of each nucleus.

The floor of each base nucleus also forms a role in this system. This is described in Chapter Four as the Harding Hive Debris Floor.

The two outer nuclei will now be called The Towers and can now have additional nuclei placed on top as high as you wish.

The queens must be given plenty of laying and storage space. Remember that no queen excluder is above the two outer tower base nuclei.

The inner nucleus is for the placement of unsealed brood; for whichever queen rearing system, either Lid or Frame method, is used.

Feeding also is directly above the queen cells.

You should now have a single story in the centre with multiple stories, known as the towers, being either side connected with 50mm pipe or larger, as you prefer.

- The pipe going through the lengthways sides of the base nuclei (see photo).

- One 50mm hole on the right hand side end of the left base nucleus (see photo).

- One 50mm hole on the left hand side end of the right hand base nuc (see photo).

- Two 50mm hole in both ends of the central base nucleus allowing for connection pipe with each outer base nucleus (see photo).

- The base nuclei have small entrances in the front.

- The outer base units also have side entrances.

- Any further nuclei put on will also have side entrances.

Clear so far?

Why Use 5 frame nucs?

✓ *It is as close to their natural environment, as I could make, re chimney or tree scenario.*

✓ *Queen cells are built more naturally as if they are in supercedure or swarming mode.*

✓ *Plenty of nurse bees in the right place being drawn from both towers.*

✓ *Good quality cells are produced.*

✓ *An abundance of royal jelly in the right area.*

✓ *It builds up quickly maintaining temperature.*

✓ It creates congestion maintaining food requirements.

✓ Easy manipulation due to its lightness in weight that makes it more manageable.

✓ Can be used for small or large operations.

✓ Queen cells can be staggered by date for a more manageable method.

✓ Various queen rearing apparatus that is available on the market can be used.

✓ It is easy to get to the centre nucleus where the queen cells are (see photo).

✓ Disturbance to the two main bodies, the towers, is limited to normal inspections.

✓ The nuclei can be split with a queen cell when queen rearing is finished.

✓ The system can be left to over winter as it stands, without feeding.

✓ It is a small enough system to possibly have close at hand ie; in your own garden.

✓ It is never queenless so never angry, unless you are doing something wrong.

✓ Queen rearing can go on for a much longer part of the season.

✓ The whole system is very easy to transport.

Why use 50mm hole (or larger) with two 6inch lengths of pipe?

✓ It allows plenty of air movement around all the nuclei.

✓ The distance between the centre base nucleus and the two towers is enough to be detached from the main body of each nest so they assume the queen is failing, therefore happy to raise queen cells.

✓ Worker bees can infiltrate all parts of the hive.

✓ Frames of bees, preferably without the queen, can be interchanged between towers.

✓ *The two queens are separated and far enough apart.*

✓ *The odour/scent is of one throughout the colony.*

✓ *If the pipe is too small, it could become blocked with propolis, and then the two towers would have separate identities.*

✓ *I used clear plastic pipe. It was interesting to watch and note what was happening.*

✓ *Drones would be content in either part of the system and not get trapped or distressed.*

Why not use full size brood boxes?

- Simply, they are just too big, plus I would rather use brood boxes for production.

- The one important factor for successful queen rearing is congestion.

- It takes too long to build up, for when I'm ready to start.

- The centre brood box would have to be decreased in size using dummy boards and insulated.

- Difficult to manipulate and would get too heavy to warrant delicate manoeuvres.

How does it work?

Depending on your resources of bees it can be started immediately; if not, then it may take you a season to build up the new colony. However you do require two queens whether you split a colony and introduce another new queen or use two nuclei that you already have.

If you split a colony then the brood must be split equally between the two outside towers, taking notice as to which side the queen is placed.

Do not worry if you cannot find her, the bees will tell you. There will be a lot of activity on the side where there is no queen, with fanning by the worker bees on the side where she is.

If your not sure check a few days later as queen cells will be started. If a queen is available introduce her to the other tower using a hair roller with tissue paper or honey candy in one end and the other end blocked up with wood or a solid material.

It is always better if your can use 2 queens which are reasonably on par with each other.

All of my nuclei they have a 68mm hole drilled through the roof, for feeding, *(I do not use inner boards with nuclei or mini nuclei)*. This is the same size as the standard screw top honey jar, if you use any other size then adjust accordingly.

This is ideal for a quick feed of diluted honey *(I don't use sugar for feeding)* approximately 60% honey and 40% hot water. Punch a few holes in the lid, easy to fill, easy to apply and normally gone overnight.

Have spare jars ready to make replacement easier.

A quick, stimulating, fix.

When not used for feeding, the jar plus lid or just the lid can be used to block the hole in the lid, everything has a double use.

Early spring

I leave my bees to over winter within this unit so build up is monitored; extra space given when needed, and feed as necessary to increase the queens egg laying rate. If one side is weak either requeen when one is available or equalise the brood between the two towers or take sealed brood from another hive to increase the worker bee population.

I try to have this unit at a peak by May 1st depending on the weather. The important part of this unit is congestion.

The strength of each tower will determine whether you take unsealed brood from the towers to place in the central nucleus.

If you cannot find the queen shake all bees off before placing in the central nucleus.

If you take from another source shake all bees off and place in central base nucleus. By shaking all the bees off you will ensure the queen is not placed in this area accidentally.

You will need two frames of food, preferably one with open pollen and one with stores, with the three frames of unsealed brood in the centre, with the frames of food on the outside position. Now close up and feed in the evening with a jar of diluted honey.

What will happen?

The unsealed frames in the central base nucleus will now be attracting young nurse bees from both towers where they will stay to feed and keep the brood warm. It will only be a matter of hours before they have there own fresh direct supply of nectar and pollen from their own front entrance.

They will never run out of nurse bees which are being drawn all the time through the tubes from both towers.

Replace the brood frames as and when they become sealed, destroying any queen cells or if you wish make up enough queenless nuclei, using any of the queen cells on the frame that look satisfactory. Alternatively place the whole frame with bees into a nucleus hive and move to a new location 3 miles away, where you can guarantee supply of your selected drones.

Maintain unsealed brood within the central base nucleus as much as possible, by doing this it keeps them in this area and it is a way of training the nurse bees to create queen cells.

What type of queen cell cups and how too use then?

Whether you use the Jenter or Cupkit plastic box method of getting eggs/larvae into the cups, or graft one day old larvae direct into the cups, make your own wax cups and graft. It does not matter which one you use they will all work. Please note in the photo, the queen cup with yellow holder, shows a day old larvae with food, that has just been grafted.

For those of you that have noticed, I have used the cup from one system and holder from another. This is just a personal preference. However the yellow holder had to be drilled out slightly larger to accept the cup.

I have used both Frame method and Lid method, both with good success using the plastic queen cell and holder system.

Frame method

Take an empty frame and fit small blocks of wood either side to support two rows of queen cells giving adequate space between each row so that each queen cell is far enough apart to eliminate burr comb.

A single row will hold 10 cell holders, a piece of wood no thicker than 3/16 x 1inch wide x the inner length of the frame will suffice and then drill 10 holes just slightly larger than the cell holder. It helps if the blocks are screwed to the frame. The row of cells can then be swivelled to make it easier to obtain the completed cells.

With some queen cup systems, thin metal holders are supplied to fit into the frame that will hold 10 cell holders. My own preference is with wood rather than metal, as the wood would not take up so much energy by the bees to warm up and will hold the temperature longer.

It is better to use the frame method when night temperature is going to be less than 18 degrees centigrade.

Once the conditions are right for you and the bees, there is no going back, regardless of family or work, so plan ahead, watch the weather forecast and pray. Oh and forget holidays!

Make up the queen cell cup frame as above and place the completed frame, with plastic cups but with no larvae as yet. Take out any frame that has little or no brood and replace with the queen raising frame in the centre of two unsealed frames in the central base nucleus.

That evening feed a jar of diluted honey. This is important for a continued supply of food to the nurse bees so they create an abundance of royal jelly.

You will know if there is a nectar flow on as they will not take the feed, so do not worry if it does not all go.

The above frame will, over the next couple of days, be prepared, cleaned and polished with a slight bead of wax around the edge of each cup, also it has the general smell of the colony so the grafts will be more readily accepted.

48 hours later go to your selected donor queen that you want to raise queens from and select a frame as below.

Once you have a good frame with plenty of day old larvae, you are now ready to graft. I use the Chinese Grafting Tool, which is cheap and perfect for the job.

Have everything ready; donor queen larvae frame, grafting to frame. While this frame is out of the nucleus a large number of nurse bees will congregate in the empty space This is perfect for when the frame of grafted larvae is put back as they will be only to keen to start work on the fresh grafted larvae.

This is how I do it.

Select a warm day with very good light, angle the young larvae frame so good light is behind you and then transfer the smallest larva you can find that has some food with it.

No bigger than these in the photo. Using your Chinese Grafting Tool or similar, graft from one cell into your prepared cell cups, as shown earlier, then replace the frame back into the queen rearing colony, feeding as required.

As I said earlier nurse bees do need training to create queen cells so don't be disappointed if you do not get 100% on your first effort. If you do get 100%, then well done!

However, if that is not the case, then check 48 hours later. If there are any empty cells, group together all the accepted cells, take away the empty cells and re graft.

The more the nurse bees do the better they get. It is possible that the larvae may have been damaged in grafting so practice does make perfect for you and the bees.

Check again 48 hours later, taking serious note as to date, how many and where the original started cells are. Don't use memory, write it down or use modern technology and take a photo and date it.

It is sometimes better that your queen cells are staggered by date. It does then make it easier when making up nuclei every day rather than all in one day.

I make up these on the 10th day of the queen cell, so you see that writing it down or taking a picture is a must and does make life so much easier. 4 days later I can then introduce my queen cell.

For protection I cut up a 1 inch length of soft hosepipe and with a small slit on one end. Place this gently over the queen cell; slit first; this will be more than adequate to protect your investment.

On that evening I feed with a jar of diluted honey, replacing when necessary. Feeding does stimulate the bees and the virgin queen to go and get mated.

Watch the weather and 2/3 weeks later pollen will be seen being carried in, a mated laying queen will be observed. To this day there is never a better feeling than seeing your rewards in this way.

The Lid Method

I am never quite sure what to call this, but Lid will do.

I have made up a clear stiff plastic cover 1/8 inch thick with a suitable wood surround to fit and cover the top area of the central base nucleus. The wood surround needs to give enough height for a full size queen cell with space below and above the frames to give movement for bees.

Too much and they will create burr comb and not enough, they will join the cells to the top of the frame, therefore damaging the cells when inspected.

The clear plastic cover is marked out and drilled to accept as many queen cells holders as you wish. The better the weather and nectar flow the more queen cells the bees can handle. I have used length strips of wood to separate each row, however it is not important. I have experimented with and without. As you will see on page 52. Within this lid, depending on size of nucleus you use, it is possible to get 5 rows of 11 cells in each row with suitable distance between each cell.

In some years depending on the weather, it is better to use this system when night temperature can be guaranteed at above 18 degrees centigrade. I have had approximately 40-ish cells in one go, when the weather was perfect, but then I was stuck for nuclei. This is how the idea originated for "The Harding Mini Nuclei Complete System", explained in Chapter Five. Having already experimented in this area, it didn't take long to make more mini nuclei.

Early in the year, or if there is a change in the weather, the outer rows are sometimes neglected. Do not worry, It all depends on how many cells you require. If you want to isolate the two outer rows it could easily be blanked off, thus conserving work for the bees. The best insulation you can have is bees; do not worry about extra man made insulation within the roof. It will just get in the way of the 68mm feed hole for the honey jar or cause unnecessary condensation that creates extra work for the bees.

The beauty of using this Lid method is that there is no disturbance to the bees. It is just a case of taking off the roof in the centre and checking each cell individually, whether you are supplying grafted queen cell holders or just checking their progress to see that all is well with your queen cells. This is so nice to do without having to put all your gear on. It is so good to be at one with your bees.

The method for the Lid method in preparation is just the same for the Frame method. Fill the lid full of empty queen cells and place above the central base nucleus. Leave for a few days while the cells are prepared by cleaning, polishing and putting a thin film of wax around the edge of each cell. By allowing them to do this, it makes it more easily accepted when you start your grafting.

Having a started wax bead around the base of a polished cell with a day old larvae and with the odour of the colony with plenty of nurse bees this will ensure better acceptance.

The plastic lid on the latest version has extra smaller access holes; I put these around the top outside edge of the lid, to help with circulation and dispersal of moist air, plus access to the feeder.

The Towers

Normal inspections are all that is required.

Once all the work has been done to produce queen cells, attention can be put towards the towers. Both towers will be working in harmony with each other, generating the same odour therefore creating no problems.

If your selected donor queen is in one of the towers then this is not a problem as the frame taken for grafting can be obtained and replaced very easily, thus conserving brood temperature.

There is no queen excluder, just like in the wild with no stress on the queen. Allowing her to lay where she wants, above the base nucleus, increasing space becomes very easy for you to supply for the queen to lay.

Just give further nucleus boxes, or it can be overcome by taking sealed brood frames and stores away to help build up other colonies. Or you can make up nuclei for your new queen cells and replace with frames of foundation or drawn comb.

The system is so versatile.

If queen cells are found (which I have not experienced in all the years I have used this system), then destroy or if it is your selected queen make up a nucleus. I tend not to waste their energy in creating a queen cell. With queen rearing you can never have enough spare equipment. At the end of your queen rearing, the system can be dismantled with the nuclei that have been used for the system can either be split to accommodate any cells that you may have left or the whole system can be left for the following season. Which is what I prefer to do.

Remember this is only part of my beekeeping puzzle; each piece is an intricate part as you will read.

Chapter Four

The Harding Hive Debris Floor

Incorporating

The Harding Hive Nucleus Floor

I am a great advocate of mesh or open floors having tried my first some 28 years ago As time went on this experimentation moved to Varroa control before eventually finding the importance of *Electromagnetic Geopathic Stress Curtain Lines in Chapter One.*

With mesh I have tried varying size holes, from using just part of the hive floor to all of it being mesh. I have experimented with no floors as it would be if they where in the wild. I found these to be very satisfactory until I needed to move them.

The first job of every season was to clear the enormous amount of debris on the floor from where they had opened stores. It was either granulated, too big for them to handle or they just could not physically handle it due to low numbers in early spring. It then lays there until we come along on the first inspection to clean it or the colony gets big enough in numbers to deal with it.

We have all seen it and more often than not it was infested with all sorts of bugs and disease bagging a free lunch. Why do we give them a solid floor when they do not have it in the wild?

So a compromise had to be found, I wanted as much of the debris to fall out free of the hive, specifically the Varroa mite, but needed the facility to move at a moment's notice, whether it is for pollination, crop or queen rearing. Easy transportation and loading was necessary.

Floors now had to be looked at in a completely different way as never before. After all, its not what we want, putting them in nice little houses, this was just to satisfy our own needs and certainly not what the bees would want!

They have the perfect places in the wild with their own microclimate that has air movement and minimal cleaning with all debris falling well out of harms way. We need them to be more comfortable and pro-active with cleansing their environment in our beehives.

It was while my children Chris and Sally (now 31 & 28 as of 2009) were very small back in approximately 1984. They were playing outside with some of my scrap sawn wood, while I was enjoying a cup of tea with my *ex wife*.

They put them parallel to each other and made a square, I am not sure what my *ex wife* made of it and my children had thought they had done something wrong, all I did was stare at it and then the light came on, it hit me.

My reaction was of sheer delight and jubilation as they had solved my problem. *My ex* could never understand why I could not switch off from the bees, and perhaps, *that is another reason why she left.*

Back to the drawing board and then into the workshop and the floor was born.

Each wooden slat was the width of the frame with 1/8-inch gap between each slat; each slat was fixed to an outer substantial frame to take stress and weight.

The distance from the bottom of the frame to the floor was only 3/8 inch leaving this distance for the entrance. Each slat had a 45-degree angle along each side of its length making sure if anything fell it would be automatically channelled out by the bees, either by physically moving or fanning.

The photo of the floor was taken on the 5th March 2009; please note how clean the floor is, after going through winter.

Benefits

- ✓ *All micro debris fell out, well away, yes, even Varroa mites.*
- ✓ *The opening is directly below each frame where the bees are working.*
- ✓ *It is as if they were in a wild environment, using underneath if it got to hot in the summer.*
- ✓ *No need for a large entrance.*
- ✓ *Less guard bees needed and better protection.*
- ✓ *No burr comb at the bottom of frames.*
- ✓ *More advanced in spring covering more frames than other colonies.*
- ✓ *Ventilation that the bees would normally control if they were in the wild.*
- ✓ *Easy transportation with no need for additional travel screens.*
- ✓ *Easy to make to fit any hive, even WBC.*
- ✓ *Much cleaner, healthier and hygienic.*
- ✓ *Nowhere for damp areas.*
- ✓ *It stopped becoming a compost heap for every bug looking for a meal.*
- ✓ *Much drier environment.*
- ✓ *Far better airflow control by the bees.*
- ✓ *Better protection control by the bees if any alien intruder tries to get in.*
- ✓ *The floor became part of the nest rather than the hive.*

Now this floor was perfect for the production colonies, which solved a major problem within my puzzle of beekeeping.

Unfortunately all of my nuclei had plywood bases and I did not want to scrap this wood so another idea was born.

The Harding Hive Nucleus Debris Floor

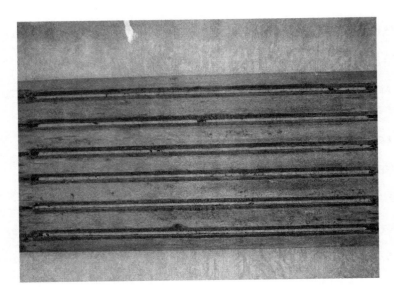

This is where health and safety might have to look away.

Each floor was unscrewed and taken off in turn. Cuts were made with an open table circular saw (the guard has been taken off), using a stacked Dado blade that is 1/8 inch wide. Great care must be taken while making these cuts.

The measurements must be accurate. Lengthwise cuts are made within the brood area of the floor, the same measurements as the hive floor. Then using a router fitted to a router table cut a 45-degree angle along each length. As with the hive floor all debris is channelled to fall out.

As there was no full cut along the base it still had rigidity with no other additional pieces added and could be re-screwed back on the nucleus.

The same benefits are assured as the Harding Hive Debris Floor.

I have felt that the bees are able to control pests, diseases and condensation far better if the debris is able to fall straight out of the hive/nucleus as it would if they where in a wild situation.

"I have not used any chemicals or sugar for more than 18 years after finding that the prescribed chemicals to treat the Varroa Mite were interfering with the health of my queens. (They died!)".

I have very little or no Varroa mites. I am working with *Electromagnetic Geopathic Stress Curtain Lines* so the bees are able to control the Varroa mite or eradicate them altogether, rather than treat with a chemical. Please refer to **Chapter One**, earlier in this book, as research will always be ongoing.

Summary

I am sure there has been very little change in practical beekeeping from books that were printed 50/100 years ago; in looking at pretty boxes or beehives that look very picturesque in using frames to inspect the bees. However, the seasons have changed since then, with bees starting earlier and finishing later with a very brief interlude for winter on many occasions in the UK. No consideration was given to the floor when the beehive was designed. This creates hurdles for the bees to overcome that do result in extra work for the bees and it usually stops debris from getting out. Perhaps the bees are now telling us that what we have put them in, they no longer like. To suggest that the bees are having extra problems with pests and diseases to deal with from when the beehive was first introduced will most probably upset a few purists of the beekeeping hierarchy.

Time moves on and so do the bees, we need to help them to, preferably without chemicals, and not hold them back. Perhaps self selection might help, allowing the weakest to die out and then breeding from the surviving fittest, as nature would want.

Help them to help you!

They live in our beehives, an environment that causes condensation and moisture that starts damp, mould, worms, slugs, and wax moth not forgetting mice.

In winter and spring the floor becomes a thriving area as a compost heap for every bug under the sun that is attracted to it and we wonder why are bees are dying, absconding or trying to cope, with disease eventually getting the better of them and that's without the contribution from The Varroa mite. Honeybees reduce in numbers for winter to survive not to become housekeepers of manmade box that looks pretty.

All colonies have a disease of some sort.

It is just the greater number of bees that help to control or hide what may be in with them.

If we can give them an earlier start during winter, which is now very short, except for the odd winter that is normal, which enables them to get rid of waste more quickly, there numbers would then be greater when it matters. Thus making it easier for them to control whatever else that may be in the hive. What we can do is give them more air circulation to dispel any condensation created so eliminate much of the above that is attracted to moisture and help to get rid of any waste.

Do not import queens. Raise good quality native queens here that are more able to deal with our climate. They are about; you just have to look and read what you see.

I know our bees have a pretty rough deal, give them a helping hand rather than trying to kill them by kindness. This dilemma of honeybees dying out did not start just a couple of years ago as highlighted by beekeepers or the media. *It started when mankind put honeybees into a log, box, skep or beehive. We are killing them by kindness and greed.* Honeybees have been dying for many years, in fact centuries, since the first man found honey and wanted to domesticate the honeybee and harvest this wonderful food, honey! *Why is there all this surprise mentality by BEEKEEPERS of Honeybees dying?*

Its time for you to listen by looking to what the bees have been trying to tell us during that time, and the bees will keep telling us. Keep looking and listening! But then who is going to listen to me if you do not listen to your own bees? We need to help our bees get back to the wild even if they are mine. Another piece of the puzzle falls into place rather nicely, read on.

Chapter Five

The Harding Mini Nucleus Complete System

I know, mini nuclei have been round for a long time so why try to improve or even standardise? Who put that mountain there if it wasn't to be climbed the most difficult way possible? I would have used a helicopter to get to the end result, because the end result is the same. It is just how much joy, effort and personal satisfaction you want to get out of it.

Here you see a super mini nuc and a brood mini nuc giving extra laying space for the new mated queen.

I have seen many different types of mini nuclei and all of them have one purpose; to supply a small portable box, so that the virgin queen can get mated in a designated area hoping for the correct drone mating. Well, that sounds good to me.

As with all my projects, I do like to get value for money in anything I make, with everything having two uses or more in life. The same applies with this system that started with the outer dimensions of the nucleus box and frame. Most of my equipment is National. This was the basis for the design for this particular system.

But as always there are questions?

- Why put a handful of bees under so much stress of building comb?

- The manufactured mini nuclei can't be used for anything else.

- Are there enough bees to sustain a virgin queen at a crucial time of her existence?

- Are there enough bees of the right age?

- Is there enough space for the queen to lay eggs?

- Can two or more mini nuclei be united?

- Can transfer of the queen be done without finding her?

- Is there enough space for food?

- Mini nuclei are a target to be robbed out.

- Is there an easy way to feed and stimulate the virgin queen to get mated?

- Is there enough ventilation?

- What happens if family, work, weather or holidays intervene?

- Weather plays a large part in the success of a mini nucleus.

The Design

The mini nucleus is 9 inches square with depth being either super or brood, as shown in the photo.

It takes 6 mini frames, which are approximately half the width, of a super frame or brood frame *(nothing new there, I hear you say)*.

The roof has a 68mm hole drilled to take a honey jar feeder, as shown. The floor is removable with a couple of screws and with a central 22mm round bottom entrance.

The debris floor also has 1/8 inch saw cuts the width of each frame and routered with 45 degree angle to dispense the debris *(as explained in The Harding Hive Debris Floor, Chapter Four)* Interchange ability was crucial to the system's success.

Additional equipment was made to ensure there was some way of getting the mini frames drawn and filled with stores and a supply of bees, so a specially designed super came into being.

The mini nucleus frame super

A super was made to take 26 mini frames with a division board in the middle; again this could also be made to brood frame size. The frames are simple and basic to make, the only part is the spacing; Hoffman, metal ends or castellated (your choice), and these frames are placed in the nuclei. It is down to personal preference if you do not wish to use any manmade spacing,

I hope that you can appreciate that it is very easy to make and assemble. Due to the frames small size, my preference is not to use any manmade spacing.

It starts by placing the super with mini frames above the brood box on a strong donor hive with the queen excluder now above this. Any additional supers can be put above the queen excluder in early spring. This gives the hive extra room and space for the queen to lay eggs into.

The following season with drawn comb, I would put this on the donor hive for wintering.

When queen cells are almost ready, prepare above hive by finding the queen just making sure she is in the brood box. Put on the queen excluder and replace mini frame super on top of the queen excluder.

You can now go to this hive a few days later and take out the frames complete with bees, with very little disturbance and knowing you do not have the queen to worry about.

Making up the mini nucleus.

A few days prior to when the queen cells are to be transferred, make up mini nuclei in the evening. Not forgetting to fill the entrance at the base with a small piece of sponge. An old honey jar lid with a few small holes is used to fill the feedhole in roof.

Take each frame with bees and place gently into the nucleus. If the frames are not all drawn then equalise them out putting frames of brood in the centre and frames with food or foundation on the outside.

These frames will now have fresh stores, pollen and brood that will strengthen the nucleus when it most needs it. The new emerging young nurse bees will help to feed and stimulate the virgin queen to get mated. I always have felt the more bees you have in your nucleus the better chance and start in life your virgin queen will have when she emerges. If you need to shake in a few more then go ahead, the more the better at this stage.

The nucleus will deplete rapidly in numbers over the coming 2/3 weeks due to drifting, dying or just getting lost. However you do have the luxury of new nurse bees emerging from the mini frames given.

The mini nucleus can now be taken to a new home, because of its rigidity and lightweight design it can be strapped, using a hive strap, to a post, tree or any upright structure that is secure and above an *Electromagnetic Geopathic Stress Curtain Line.*

I have made a two tier wooden structure that will hold 10 mini nuclei; you may want to do the same. The purpose was to take advantage of the bottom entrance. Take out jar lid in roof and replace with a jar of diluted honey. This is given to help calm the bees and make sure it puts them in the right mood to accept the queen cell.

Do not forget to take the sponge out of the bottom entrance after transporting. I always put some grass in the entrance to stop too many bees coming out and getting lost.

Within a few hours after dusk this will have dried and fallen out with the activity made by the bees so when they emerge in daylight hours reorientation will take place.

On about the 14th day of the queen cell, transfer to a mini nucleus using a soft hosepipe protector (cut hosepipe into 1inch length and cut one end so that it will then gently slide over the queen cell, as shown below).

There is no need to take off the roof just take off the feed jar and gently place the protected cell in between two frames, causing as little disturbance as possible.

Now leave and pray that the weather holds out, and always making sure there is feed available.

For the past two years the weather has changed for the worst on May 15th 2007 and May 14th 2008, so 2009????????????? I don't know.

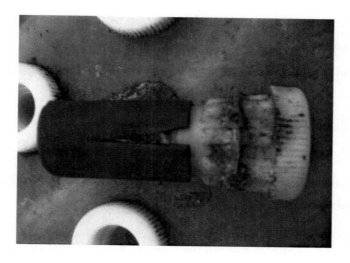

Two to three weeks later, pollen will be seen going in which is a very good sign and your excitement to look in will be overwhelming… Finding a new mated laying queen, what a site to behold! It makes everything worthwhile knowing that you are securing next years production queens and sustaining or improving your own strain.

Now you have got your newly mated queens so what are you going to do with them?

I do not like using any queen in production colonies until I have seen her brood and offspring. However that does create a problem…a young mated queen clambering for space to lay eggs…If I leave her the bees may decide her future or swarm, resulting in all the good work going out of the window.

What I need is to keep her stimulated in laying eggs with extra frame space. She is now going to help me produce the next generation of mini nuclei. I did say earlier that the base comes off with a couple of screws so another mini nucleus is placed on top, with either drawn or foundation frames.

Using extra space to delay when you want to use the newly mated queen, and then keep going with more mini nuclei, feeding as an when necessary or as the weather dictates.

When you have decided queen "A`s" future you now have 2/3/4 mini nuclei ready for more queen cells unless you want to unite these now queenless nuclei with a queenrite "B" nucleus. Thus delaying queen "B`s" future until you are ready.

This system is so flexible with what you want to do and when you want to do it, in case time, weather or family is not on your side.

Finding the queen can be difficult for some individuals and playing hard to get on other occasions but you may want to transfer on to normal size brood frames getting the new queen ready for the production colonies. It is simple to do.

- Take the floor off two mini nuclei, one queenrite and one queenless or empty with frames.

- Place both mini nucs on top of a 9-inch width nucleus with full size national frames.

- The bees will soon unite and use the brood frames.

- The only way out is through the entrance of the larger nucleus.

- When you see the queen laying on the brood frames then the mini nuclei can be recycled elsewhere.

Queens on tap so to speak, without the rush or panic, well not much and even if the weather changes, the extra space given will keep them happy for a few weeks.

Now you have a system, which you can work around the family, work, holidays and the weather.

The entrance, why in the base and why 22mm?

The original idea of 9inches square was for 4 mini nuclei to fit on top of a National hive with the entrances being at the front of each mini nucleus and facing a different direction. The heat from the hive below would help the mini nuclei control temperature. However things evolve and get changed with different experiments.

It happened in this case with the following.

Any mini nucleus system is under threat of attack from predators when the entrance is used at the front. Many hours have been spent watching and cringing at the loss of another good queen being destroyed by these attackers. Until one day, by accident, on one of my earlier experiments, it had fallen off its stand and had been in this position, on its front, for a few weeks with the entrance now underneath but still off the ground. Rather than move it I watched it over a period of time. To my surprise the defence mechanism was far better when predators were about. Whereas the front entrance nuclei were well under attack, this nucleus was handling the problem keeping their attackers at bay. It was stronger in brood and food due to possibly having less stress within. So the following year I introduced base entrances in all mini nuclei with not one being attacked or robbed out.

THE ABOVE WAS A READ WHAT YOU SEE MOMENT!

I experimented with varying size entrances in the base from 15mm up to 68mm. The reason I chose 22mm, was that the larger the entrance the more guard bees were needed. When it was smaller it was propolised so 22mm was the perfect size for guard bees to protect.

Where you have mini nuclei in close proximity, worker bees and also queens on their mating flight, can be distracted. A form of identification was needed. Having dealt with electrics, car mechanics, woodworking and plumbing over the years, there was always something lying around so, 22mm white plastic plumbing pipe was always in the garage. This was used and cut to varying lengths with different colour tape being put around it.

This then fits neatly in the base hole and solves the problem of drifting, giving further protection and identification.

A similar design Harding Hive Debris Floor is used in the mini nucleus system, as in the nucleus and hive floor that becomes part of the nest, which is explained as The Harding Hive Debris Floor, in Chapter Four.

When you or the season has finished and all the new queens have been dispersed the frames complete with bees and brood can be put into your mini frame super. This being placed back above the brood box and below the queen excluder of your donor colony, ready for next spring.

Or you can merge as many mini nuclei as you wish, to over winter. Which I have done, very successfully, as shown in the photo. Please note hive strap holding mini nucleus to concrete post which is directly above an *Electromagnetic Geopathic Stress Curtain Line.*

By using the old queens from any exchanges done from that year, their job is done but can be held in abeyance in these smaller hives just in case they are needed in an emergency.

They have just as much room as a standard national brood if they are stacked in this way and if there are any failures, they will be there ready for you at anytime. Also more importantly, they will be there next spring continuing the mini nuc system then being replaced by next years virgin queens, consequently keeping the cycle going as the year goes on.

I prefer to over winter the production colonies with a minimum of two brood boxes with one super or mini nucleus frames super giving the bees plenty of storage and expansion in spring.

Depending on the queen rearing, which will determine, if I split the double brood in that season, increasing my stocks yet again.

That is the unknown, as we all know. Nothing can be perfectly planned with beekeeping. You have to adapt to read what you see in front of you, having spare equipment is a must, just in case.

I have said before the best insulation is bees but with plenty of ventilation. This cuts down on condensation and moisture and creates a dryer atmosphere. *(It is what they would have in their wild environment).*

If they go into winter strong then it makes spring easier, and with the help of what brood and food is left in the mini nuclei super, this gives a boost at a much needed time and it will give space for winter stores.

Benefits

- ✓ *Easy to transport due to symmetrical size and design.*
- ✓ *3 or 4 mini nuclei can be carried at once using a hive strap.*
- ✓ *Easy debris clearance using the Harding Hive Debris Floor.*
- ✓ *Plenty of ventilation.*
- ✓ *Easy to produce frames with food and brood using the modified super.*
- ✓ *Less stress on the bees used.*
- ✓ *Ease at putting in queen cell.*
- ✓ *Bees have less work to do in building comb, so.*
- ✓ *More attention will be given to the virgin.*
- ✓ *Nuclei easy to make up with frames and bees.*
- ✓ *Quick and easy way for the virgin and bees to identify their own 22mm entrance.*
- ✓ *Extra space can be given when needed.*
- ✓ *Delay in using any mated queen if you're not ready.*
- ✓ *Easy to feed.*
- ✓ *Less likely to get robbed out.*
- ✓ *The system can work round family, work and holidays.*
- ✓ *Everything is interchangeable, except the brood and super frame.*
- ✓ *Great way of storing queens till your ready.*
- ✓ *Equipment is not left idle and can be used all year.*
- ✓ *It can be left in use over winter until needed in spring.*
- ✓ *It's also a perfect size hive to train a child.*

The above is a complete mini nucleus hive system.

Happy Beekeeping.

Dear Reader,

Well that's it, I hope that I have given you something to consider that is thought provoking. A combination of circumstances where beekeeping has changed my life, cost me my marriage and strangely the position of where I sleep more important a new way of thinking in how I keep my bees. I can only go by my own life puzzle of beekeeping, which will go on for a lot longer than me, my ex wife Cynthia loved doing puzzles. However it has been great fun putting the pieces together and I would not have had it any other way, well except getting divorced. I hope you have enjoyed this book as much as I have had in writing it. I am very glad to share my life experiences with you.

I've been intending to do it for years.

I am not educated, or a scholar, scientist, professor, doctorate, pharmacist or a writer, I am just a passionate beekeeper that does not want his bees to die.

By the way the quote of Albert Einstein is a misquote from Albert N Stein.

For Now, The End! Or is it?

Thank You for reading my book,

John Harding

PS,

I have mentioned my ex wife Cynthia, while typing this book, she did have a lot to put up with during our 30 years together, it was not her fault for my passion for beekeeping, it was just unfortunate that she was allergic to bee stings, the adrenalin syringes are still in the fridge, just in case.

There is one point that I would like to make that Rolf said to me after the divorce, so far he hasn't been wrong.

"The reason and cause of our marriage break down happened over a few years, just like so many marriages. Rolf can and did a long distance dowse, he was able to pick up on two strong Geopathic Stress Lines going under our bed, one going through my head, which I did suffer terrible head pains and the other going through the abominable area of my ex causing miscarriage and ongoing problems. He was quite adamant these lines were the cause and reason for our ill health and ultimate divorce due to how long we had slept in that same bed position above these Geopathic Stress Lines"

Rolf was unaware at the time of my ex or my symptoms.

I tested the area myself and he was spot on, it really did scare me because of his accuracy of where he said the lines would be, my only regret, is that I did not know Rolf sooner. I have since moved my bed to a line free area and pleased to say I have had no head pain or sleeping problem since, unfortunately it is too late to change the divorce. However, I only wish that I may get the chance to explain to my ex wife, one day, who knows perhaps this book may help, we'll see.

Thank you for our children and 34 years, my lovely, I really do miss you. (Closure)

"When anyone now tell me of illness or otherwise I can always guarantee a Geopathic Stress Line under the house/bed is involved, when tested is normally proved".

Further reading for Geopathic Stress Lines.

There are many illnesses that are affected by Geopathic Stress Lines; here are just a few that you will read about in Rolf's book;

Cancer
ME
Kidney
Liver
Heart trouble
Stomach complaints
Insomnia
Rheumatism
Ear and teeth disorders
Leukaemia
Emotional and Mental conditions
Period pains
Peaceful sleeping
Psoriasis

Our health is harmfully affected by Geopathic Stress Lines.

Ask yourself?

Why does the UK have the highest Cancer patients in the world?

Nature's answer is there. The medical profession are too blinkered to see it, but then you would rather put your trust in your Doctor giving you a pill or painful treatment. Unfortunately when a mistake is made the patient dies and become a statistic.

Plants and Trees are also affected by this phenomenon. Don't forget Oak trees and Elderberry mentioned earlier. Knowing where to plant may help the;

Orchard Growers of top fruit.

Pears, Apples, Plums would all benefit.

Horticultural growers.

Rose growers.

Vegetables.

Fruit bushes.

Even to find the best place for a compost heap.

Animals too, are affected as they can sense it;

Cats (will always sleep on a GSL).

Horses.

Sheep.

Cows.

Pigs.

Nesting Birds.

Fish in Aquariums.

Chickens.

Basically most things are affected one way or the other. It is how nature works, being drawn to it or repelled from it. Understanding it is the first problem... after that everything makes a great deal more sense, as to why some things are as they are.

Just by a simple check and moving your bed could help you avoiding Cancer or related illnesses and will give you a healthier life style. It costs you nothing, and could save the National Health Service millions of pounds. It has got to be worth it, as you have nothing to lose except your life! Do it, try it today, it will pay dividends for you in. *Living longer!*

Don't see what you have read,

Read what you see!

If you feel the need to contact me then please do, at the following;

harding@clavies.freeserve.co.uk

01384 423557

Or

07974121472

"READ WHAT YOU SEE, DON'T SEE WHAT YOU HAVE READ"

The future,

To save

"The Honeybee"

Is in your hands

Right now!

An HOLISTIC Way In Saving The "Honeybee"

By John Harding

Lightning Source UK Ltd.
Milton Keynes UK
28 February 2011

168390UK00001B/5/P